The All Hearts Club

(Book 1)

All Hearts Club

Friendship Club

Deborah Waterhouse

Acknowledgement

To: Julie Dulaney Bolt and Shellie Wyzard Nall, the two best friends a girl could have. Your friendship has been a true gift from God and I love you both so much. Thank you for a lifetime of wonderful memories. You both mean the world to me.

To my children: Evan, Emily, Lauren and Rob, I hope everyday you know how much you are loved. Always chase every dream and never settle for anything less. To my grandbabies: Morgan, Joseph, Mason, Harleen, Parker and Bentley, I love you all the way up to the sky, $5 and a penny. You all are my whole world.

To my family: Mom, Dad, Billy, Dorothy , Brandon, Bailey, Matthew, Barrett,Theresa, Pat, Kyle, Katie, Eli, Charlie, Sadie, Lou, Allie, Tim, Holly, Michael, Cora and Liam I love you all so much and thank you for everything you have done for me. To my Stony Point family: thank you for all your love and support and giving me a firm foundation in God's teachings and being a wonderful example of his love. To my Hobby Lobby family: thank you so much for your love and support and laughter.

You guys are AWESOME. Who knows Jack I may just have to pass on my rock on to you. To Wanda, Ann and Layne thank you for the love, laughter and friendship. And most importantly, I thank God, for without him I would be nothing.

" Time to get up, my sweet Ellie. It's your first day at your new school." Momma said sweetly. Ellie gave a small moan as she got up and got ready for school. She was very nervous. What if the kids didn't like her? Ellie really missed her old friends and her old school.

At breakfast Ellie fiddled with her flapjacks. She was very nervous. " Momma, I have a tummy ache, maybe I should stay home today." Ellie said softly. "My sweet Ellie, you are just nervous about your first day, but don't worry my sweet,smart girl you are going to love your new school, and you are going to make lots of new friends." As Ellie finished her breakfast, her tummy tumbled and turned with the first day jitters.

"Ellie, time for the bus." Momma said as she handed Ellie her backpack and gave her a hug and a kiss. "You are going to have a great day. Remember Mrs. Gazelle said that today was a very special day at school because today is club sign up day." This made Ellie very excited.

As the bus pulled up to Ellie's stop, Ellie's tummy began to tumble again. "Good morning, you must be Ellie our new student. I am Mr. Dingo. Please find a seat". " Nice to meet you Mr. Dingo." Ellie said shyly. Ellie looked for a seat.

First Ellie tried to sit with Bella but Bella said "NO! only beautiful girls can sit with me."

Next Ellie tried to sit next to Braxton " NO GIRLS!!!" He grumbled.

There was no room in Ginny's seat. She had too many books.

Marko and Arnie turned their heads away, and Ellie passed by them.

Sam and Ally whispered and giggled as Ellie got to their seat.

Finally, Ellie found an empty seat and slumped down in it. A tear rolled down her cheek as she softly said to herself "I miss my old friends and my old school and my old bus." As the bus headed on towards school Ellie thought, "Oh, this is going to be a horrible day."

Ellie was so sad that she didn't see Annie sitting in the seat across from her. Annie wanted to say hello to Ellie but she was very shy.

At school, Ellie walked very slowly into her classroom.
Again, all the kids whispered and giggled as Ellie walked by.
"This is a horrible day." Ellie thought to herself.

Mrs. Gazelle gave Ellie a warm welcome. "Good morning children, I want you to say hello to our new student Ellie. Let's make her feel welcome. Ellie you may take your seat next to Annie.

The children still whispered and giggled as Ellie took her seat. "This is a HORRIBLE day." She thought.

At lunchtime, Mrs. Gazelle announced, "Now remember children after lunch we will be having Club signups so be thinking about what club you want to join. Enjoy your lunch."

Lunch time wasn't any better for Ellie. Her milk had spilled in her lunch box and her lunch was a soggy sloppy mess and all the kids ignored poor Ellie. "This is a terrible horrible day!!!

After lunch Mrs. Gazelle announced "All right children, it is Club signup time..." Ellie got up slowly.

SPORT CLUB

Braxton and Sam had a sports club, but Ellie did not know much about sports so she passed by them.

Ginny's Club was a space club, but Ellie did not know much about space so she passed by.

SPACE CLUB

GYMNASTICS CLUB

Marko and Arnie had a gymnastics club, but Ellie did not know much about gymnastics so she passed by.

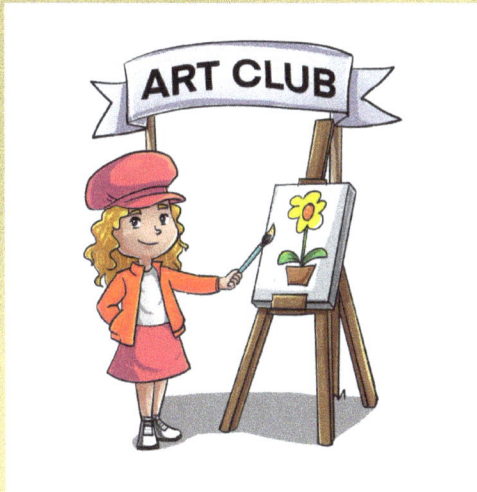

Ally had an art club, but Ellie didn't think that she could be a good artist like Ally, so she passed by.

Bella had a beauty club, but Ellie didn't think that she was pretty enough so she passed by.

Ellie was very sad. She didn't fit in anywhere.

All Hearts Club

Just then Ellie saw Annie's Club, it was the "All Heart's Club". "What is the All Heart's Club?" Ellie asked quietly. "It is a friendship club and anyone that has a heart can be in it. Everyone is welcome in my club." Annie replied. Ellie was very surprised. "You want me to join your Club?" Annie smiled "Yes I do. I think you are perfect. You will make a great friend." Ellie was very happy. "Yes I would love to be in your All Heart's Club Annie."

Friendship Club

At recess Ellie and Annie began their very first meeting of the All Heart's Club. First, Ellie and Annie drew pictures with chalk on the sidewalk.

"Those are great pictures. Can I join your club?" Ally asked. "Yes everyone is welcome in the All Heart's Club." And the girls all drew together.

Next Ellie, Annie and Ally went to walk on the balance beam by the jungle gym. There were lots of giggles as the girls wobbled back and forth as they walked on the beam. Seeing the girls laughing and having so much fun Marko and Arnie came over. "Cool club, can we join?" "Yes everyone is welcome in the All Heart's Club." And they all took turns walking on the beam.

Next, the club members ran around pretending to be airplanes. They zoomed. They did spins. They soared, but most importantly they laughed. "Can I join your club? I love to play airplanes." Ginny asked. "Yes everyone is welcome in the All Heart's Club." And they soared all over the playground.

Marko picked up a ball and asked the kids if they wanted to play kickball. "Yes" they all said cheerfully and begin to play and laugh. Soon Braxton and Sam came over "Can we join your club? We love kickball." "Of course, EVERYONE is welcome in the All Heart's Club." It was the best kickball game ever.

As they played Ellie saw Bella sitting by herself. Bella was very sad. "What's wrong Bella?" Ellie asked. Bella wiped a tear and said "nobody wants to be in my club."

Ellie put her hand on Bella's hand "well you can be in our club. It's the All Heart's Club and everyone is welcome to join." "I can?" Bella was very surprised. Ellie and Bella went to play with all the other members of the All Heart's Club.

All the children laughed as they all played together.

Mrs. Gazelle called for the children to come in from recess.

After the children all took their seats, Mrs. Gazelle asked "So children, what club did everyone join?" All together the children announced "The All Heart's Club." "What is that?" Mrs. Gazelle asked. "It is a club for friends. Anyone that has a heart is welcome in our club." Ellie said with the biggest smile. "I think that is a fantastic Club to be in. I'm so proud of all of you. It is very important to have friends." Mrs. Gazelle said.

Ellie was so proud to be in the
best club of all. She smiled and thought
to herself "This is the best day EVER."

On the bus ride home, Mr. Dingo couldn't help but giggle as he listened to the children talk about the All Heart's Club. The children laughed and talked about all they had done at the very first club meeting and made plans of what they would do the next day.

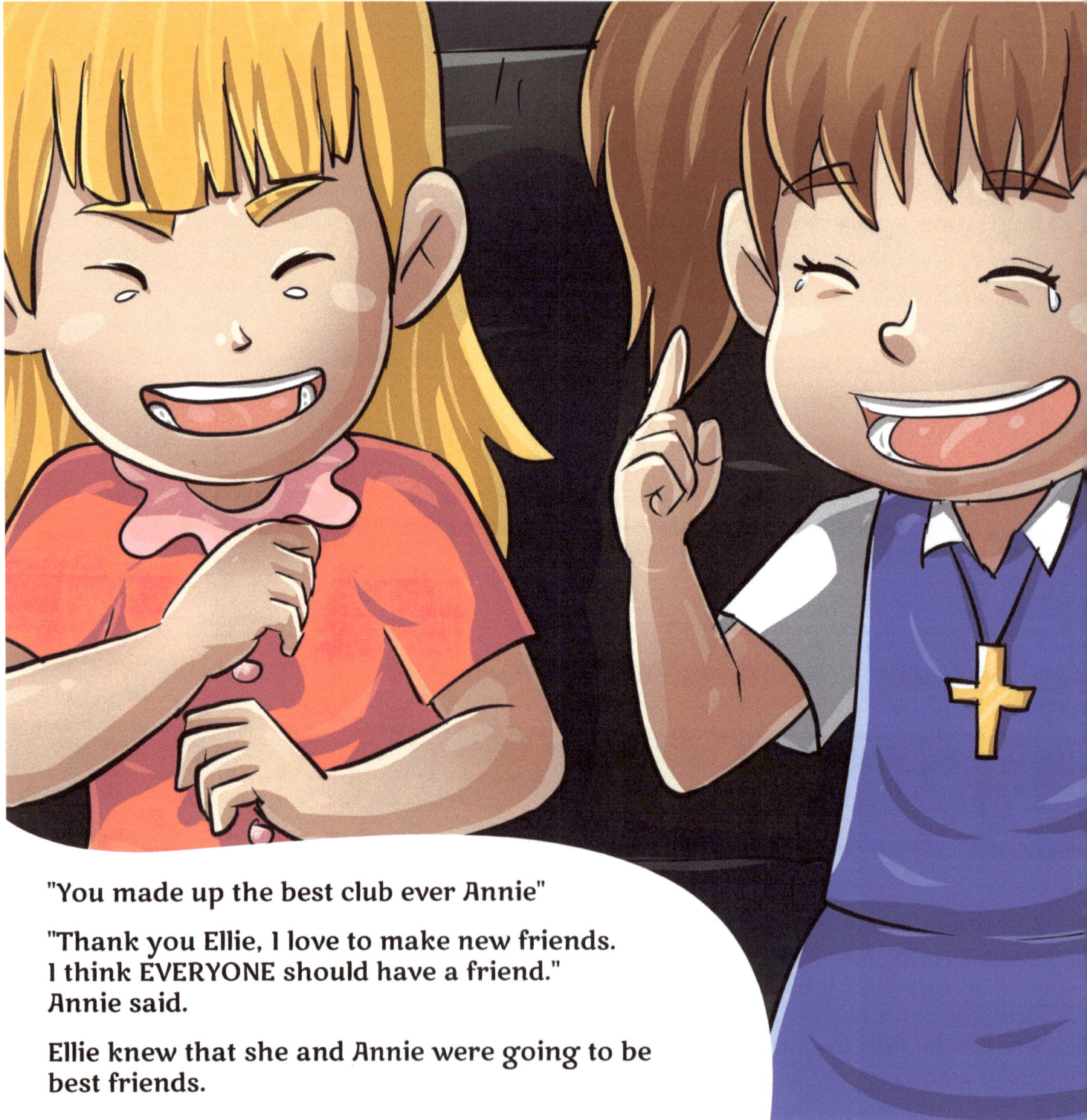

"You made up the best club ever Annie"

"Thank you Ellie, I love to make new friends.
I think EVERYONE should have a friend."
Annie said.

Ellie knew that she and Annie were going to be
best friends.

When Ellie got home she gave her momma the biggest hug ever.

"How was your day my sweet girl?" Momma asked Ellie. "It was the BEST DAY EVER. I join the All Heart's Club, it's a club for friends and everyone is welcome in our club. We laughed. Drew pictures. We did gymnastics. We played airplanes. We played kickball. But most importantly we made everyone happy. I made lots of new friends but Annie is my new best friend. I can't wait to go back tomorrow." Ellie was very excited and all of her new day jitters were gone. She was very happy.

"I knew that you would have a great day my sweet girl." Momma said as she gave Ellie a kiss on her forehead.

As Ellie laid in bed that night she
thought about all that had happened.
She had made lots of new friends and was in the best club in the world.
"Yes this was the BEST DAY EVER." That night Ellie had the sweetest dreams.